Nosebleeds, Heart Murmurs,

And Llamas

Stories from a Clinic

Jeffrey Blake, M.D.

To those who teach and mentor

"The good physician treats the disease; the great physician treats the patient who has the disease."

Sir William Osler

Contents:

Preface

VIM Berkshires is a free clinic for patients of limited means without health insurance and access to health care. It is based on a volunteer model started in Hilton Head, South Carolina in 1992. There are now over 100 such clinics around the country. Volunteers in Medicine (VIM) Berkshires is the only one in Massachusetts. Although based on the same model, each VIM clinic is different. Size, services offered, and organization are all determined by the specific local VIM clinic.

VIM Berkshires opened its doors late in 2004. In 2005 we had a limited number of volunteers and only a few hundred patient visits. The clinic has grown. Now there are over 4000 patient visits per year. We have over forty clinical volunteers including doctors, dentists, hygienists, mental health therapists, nutritionists, acupuncturists, social workers, massage therapists, and Spanish interpreters. We have many more administrative volunteers. All of these volunteers are supported by a dedicated, committed staff.

VIM Berkshires is a special place. The volunteers, staff, and most importantly the patients, sense how unique it is. As a volunteer doctor I hear patients express their gratitude on almost every visit.

The atmosphere in VIM Berkshires radiates hope and compassion.

Those who founded the clinic back in 2004 always said they would be happy to close the doors if the need went away. There was some inkling of that back in 2006 and again in 2010 with the advent of Romneycare in Massachusetts, then Obamacare nationally. Those programs do offer better access to health care but they leave gaps. There are many people in Berkshire County who have limited means, whose lives teeter on the edge, who lose health coverage between jobs, who never signed up for available health insurance, or who are new immigrants. They work in our restaurants, clean our houses, do our landscaping, work on our farms and on our construction crews. These are the people who are patients at VIM Berkshires. We care for them, bind their wounds, and try to get them enrolled in coverage. But many fall back into the cracks, the gaps. They get sick. The need for VIM Berkshires continues to grow. It may grow even more after recent election results.

This book is a compilation of stories about some of the patients I have known over the last twelve years. I cannot help but be impressed by the humanity and grit our patients show in the face of their struggles. As I learned way back in medical school, patients can teach us so much.

Leonard Cohen writes in one of his songs,

"There is a crack in everything. That's how the light gets in."

Despite all the cracks in our health care system, our patients know that VIM Berkshires is a place where the light gets in.

Note regarding patients:

The stories in this collection reflect episodes about real patients. Care has been taken not to use patient names or specific identifying factors. In the one story where a name is used, that name is fictitious. Some of the stories describe unique and unusual patient presentations. Even though their names are not used, these stories might possibly make those patients identifiable. In such instances the patients gave their consent to have their VIM medical story used in this collection. I express my sincere appreciation to them.

Dialing for Doctors

I am on the patio intent on grilling a fresh cut of halibut. The breeze carries a soft hint of late-blooming Miss Kim lilac from near the garage. The kitchen phone rings.

"It's for you. Some doctor wants to talk with you. I didn't recognize the name."

She opens the door and hands me the phone.

"This is Matt Mandel. I'm a retired anesthesiologist. Got a few minutes to talk?"

And so it began. He outlined an idea for a free clinic for uninsured people that he and some others wanted to start here in the Berkshires. Wary, but curious, I listened until he paused.

"How did you find me?"

"Oh, someone in town said there was a retired cardiologist recently moved to Sheffield. I did some checking and decided to give you a call."

Mandel and others were forming an exploratory committee. Would I be interested in attending a meeting next week in Stockbridge? I asked a few more questions, like who else was involved and how far along they were. Thinking it can't hurt to hear them out; I agreed to attend the meeting. Little did I know what was in store.

That was the late spring of 2003. Much organizational work followed. Volunteers in Medicine (VIM) Berkshires opened its doors in the fall of 2004. Back then I knew very little about primary care medicine or the uninsured population in our county. Over twelve years I have learned a great deal on both fronts. I think I have evolved into a better doctor.

And Matt, the anesthesiologist? No, he does not see patients. But he has been on the Board from the start, watching and cajoling as VIM evolved from an idea into a busy primary care clinic. We continue to need more volunteer clinicians, more specialty sources in the county for patient referrals, and more donations to keep the doors open. How many calls has Matt made over the years to achieve those ends? I have no idea. But I know one thing. He is still dialing.

Llama Man

I found it incongruous and I think he did too. On one of his first clinic visits I had asked what kind of work he did. He took care of llamas, worked for a family that raised them. Why did I find it incongruous? He was a burly man in his 40s and had immigrated to the U.S. from Russia. How does someone who leaves the Ukraine end up in the Berkshires caring for llamas? Well, that was just the job he found after coming over here. He had not been a farmer. I think he worked as a policeman in his home country but he would never say for sure.

It became a joke between us. On his visits I would always ask how the llamas were doing. He would get a mock serious look on his face and tell me llamas were nasty animals. He said that he didn't like them and only did the work because it paid well. His medical issue was high blood pressure, fairly severe. Controlling it required several medications. But over time we got there. He made strides with his diet and stopped using alcohol. Early in his time at the clinic his English was spotty but after two to three years he was fluent. We began to talk about other things.

"You know my wife is getting fat. She likes American food." I suggested he bring her to the clinic and let us check her out. She was overweight and her blood cholesterol levels were sky high. We never did make a lot of progress with her weight, but did get her cholesterol numbers down.

He was very proud of his son. The kid played soccer on his local public school team. On one visit the first thing the Llama Man said after he entered the room and sat down was "Two goals." It took me a second but I knew he meant his son's latest game.

I found out he had lived and worked in a small city near Kiev. I once asked him about Chernobyl. He was a young man in 1986. All he would say was that he was not exposed but that it was very bad and the government told them very little.

Three years ago he came in and told me he was about to get his American citizenship. He said this with great pride. I congratulated him. I then asked if he would still take care of llamas as a U.S. citizen. He put on his mock scowl and said yes, at least for a while.

Along with his citizenship came access to the American health insurance system. He went on to get Massachusetts health insurance and his care was transferred to a private primary care doctor in Great Barrington. He would make that move with all his primary care preventive issues addressed and his blood pressure under good control. We clinic volunteers like seeing our patients "graduate."

I must admit I miss the Llama Man. I sure would like to get his take on current day Russian political machinations vis-a´-vis our newly elected president and the Russian Premier.

A Kiss in the Night

Most of the time routine follow up visits are just that – routine. Often you have time to explore non-medical aspects of a patient's life. And in those comfortable exchanges you sometimes detect a factor relevant to the patient's health. The young man sitting in the exam room and I go back several years. He's burly and stocky, having gained some weight and muscle from when we first met. During his ten years in the States he has become fluent in English. He now has three sons. I like his open and steady gaze.

Midway through the visit, as I am about to begin the exam he says, "I'm famous, you know."

He has a shy smile on his face, not offering more. I take the bait and ask what he did to become famous.

"They wrote about my case in some medical journal."

I pause. Bells go off in my head.

"You mean the doctors in Boston?"

This saga starts four years ago. Our clinic handles mostly routine primary care and preventive medical issues. But sometimes pounding hoof beats do not mean horses. They can herald the unusual and strange, like a zebra. Let me paint the picture of our "famous" patient.

It was an evening in early September, almost dark. He came to the clinic after finishing work as a laborer on a dairy farm. I had seen him once or twice before for minor concerns. On this visit he was worried because he had fainted while playing soccer two days prior. His brother said he had been out about 15 seconds. He felt fine immediately afterwards. His general health had always been good. There were no findings on his physical or cardiac examination. We thought maybe he had fainted from dehydration – it had been a hot day. But his EKG had a mild irregularity. Because of that finding we arranged some tests, one of which was to have him wear a continuous cardiac monitor for two weeks. A week later, also while playing soccer, he fainted again. The monitor recording showed ventricular tachycardia – a life threatening heart rhythm disturbance. Paramedics responded and transported him to a local hospital. Extensive testing was done and his treatment included placement of a pacemaker and internal defibrillator. The situation improved but was not fully controlled. He ended up referred to a major teaching hospital in Boston where the diagnosis of Chagas Heart Disease was made and then finally treated under the guidance of the Centers for Disease Control in Atlanta. Chagas Disease is caused by a parasite named after the Brazilian doctor who discovered it. The parasite lodges in cardiac tissue and slowly causes damage over a period of many years – usually not becoming apparent until 20 or more years after the initial infection. The disease is rare in the U.S. because the parasite is not found here.

Doctors and medical students who peruse the *New England Journal of Medicine* can now learn about Chagas Heart Disease in the U.S. by reading about our patient from rural western Massachusetts. He *is* famous.

The "I'm famous, you know" visit occurred two years ago. Our young man is now approaching 40. He remains well, still works long hours at the dairy farm, and still brings his sly sense of humor to clinic visits.

We sometimes talk about his childhood in rural Mexico. He tells me how poor his family was. He and his brother slept on dirt floors in an adobe-walled one-room house. The Chagas Disease parasite, Trypanosoma Cruzi, is transmitted by the bite of an insect. The insect is common in poor rural areas of Central America. It lives in dirt and also in the mud cracks of adobe walls. It bites at night, usually while people sleep. For some reason it frequently bites on the lips, hence its nickname, the "kissing bug." Unlike others stolen in the night, a kiss from this visitor is far from benign.

Guatemalan Grit

Our bilingual medical assistant ushers in the next patient. She has done the intake and vital signs. She will also serve as the medical interpreter for the history and physical exam if needed. The patient and I greet each other.

"I want to practice my English," he says. I have a little medical Spanish and I admire his determination to better his English skills. I look at Gladis. She nods, so we give it a shot without her.

He had come to the clinic a month ago for a severe contact dermatitis – probably poison ivy. He works as a landscaper. The dermatitis has resolved. He now returns for a full preventive medical evaluation. Like many of our patients he is a new immigrant to this country without access to health care.

"What is your home country?"

"Guatemala."

"Any family here?"

"No. I have a wife and daughter in Guatemala."

"When did you last see them?"

"Two years ago when I left home."

He is thirty years old and says he has always been strong and healthy. He has never seen a doctor before coming to the U.S. I review his health habits, confirm that he has never had immunizations and begin the medical review of systems looking for something that might point to possible issues. I do not expect to find much and all my queries elicit nothing. I have one last question before beginning the physical examination.

"Any pain or discomfort in your muscles or joints?"

"I get needles in my knees."

I am not sure what to make of this. Does he mean numbness or sharp pain? He repeats the word needles. I decide we need the intervention of our medical interpreter and call her back into the room.

"Tell me about your knees. Are you having pain? Any accidents or injuries to your knees?"

"I have needles in my knees. I fell into some cactus needles in the desert."

I eye him and Gladis quizzically. She takes charge. There is a long rapid exchange in Spanish leaving me in the dust. She summarizes after a few minutes. To enter the U.S. he spent most of one night running through the desert in Mexico and then across the border. He fell several times. It was dark. Some of those falls were into cacti and some of the cactus needles pierced the skin of his knees and legs. Pieces of cactus needle had broken off and left fragments under the skin. Two months ago a piece of needle had worked its way to the surface of his skin and he had pulled it out. He wonders if he has more cactus needle fragments under his skin and if they might cause future issues like infection.

His full medical exam is normal, including the exam of his skin and knees. I detect no evidence of additional needle fragments under the skin. We set him up for tetanus and hepatitis vaccinations. He is a polite, soft spoken young man. He is sending much of his paycheck home to his family in Guatemala. In his two years in the U.S. he has learned English pretty well.

It is hard not to admire the grit of these people when you see it firsthand. I want to know more about his story, like how he ended up in Great Barrington, Massachusetts. But there are limits to the probing permitted during a medical encounter and there are other patients to see. I wish him well and remind him that he can return any time if he has other issues. Maybe then I might learn more about night time border crossings and his journey from the Arizona desert to western Massachusetts.

Shoeless

The patient is tall, thin; his face angular. He moves with considered slowness. He looks depressed. The intake form says back pain. The vital signs are normal. Via the interpreter we establish that he is from El Salvador, been in the U.S. about a year working odd jobs as a housecleaner and as an after-hours restaurant cleaner.

His responses to my queries are hesitant, his eyes shy. It is not just his back that hurts but his ribs, shoulder and elbow. He finally admits that he was mugged in Pittsfield last week and wants to make sure he is okay. He had been punched and knocked down. His wallet and jacket had been stolen. The mugging occurred on his way home from his late night restaurant job.

"They even took my shoes." He is wearing a battered, torn pair of sneakers.

He reported it to the police but has been told not to expect much. He shares an apartment with a Salvadoran friend. His family remains in El Salvador. You have to feel for the guy.

A careful exam shows some contusions and abrasions but nothing more serious. I tell him I do not think he has any broken bones and does not need x-rays. I schedule him to come back for a more complete preventive exam. We also offer him mental health counselling and suggest some contacts where he can get free shoes and a new jacket. His spirits seem to lift a little. I think the mugging hurt his pride and sense of self more than it hurt his body.

He returns a month later for the full preventive exam. His demeanor is lighter, his movements less hesitant. Because he is over 50 we examine his prostate to check for possible issues. He has no symptoms. To my chagrin he has a hardened nodule in the left part of the gland; almost certainly prostate cancer. How am I going to tell him about this?

I step out as he gets dressed and bring the interpreter back in. Just as the patient is coming back from the blow of the mugging, I now have to add to his woes. I deliver the news as gently as I can, trying to put a positive spin on it. We set up the Urology referral.

Thinking of his stolen shoes my mind somehow wanders to the legendary baseball player, "Shoeless" Joe Jackson. Back in 1920 little kids famously cried "Say it ain't so, Joe!" when he was barred from baseball for allegedly cheating in the World Series. This is just what I want right now; to tell the patient "it ain't so." But I can't do that. Delivering bad news is a necessary, but never pleasant, part of doctoring.

This story has a more positive update. Our shoeless patient did have prostate cancer, but not an overly aggressive type. He has had treatment and is doing fine.

Blood Pressure Barometer

We measure blood pressure with an instrument called a sphygmomanometer. The cuff inflates around the upper arm and pressure is measured as the cuff is slowly deflated. Blood refills the arterial system that had been compressed creating a pressure wave with each heartbeat. Simple physics. The level of those pressures, both systolic and diastolic, tells us a lot about the condition of a person's vascular system.

In one particular patient I could usually tell if his blood pressure was high by the color in his cheeks. He worked as a landscaper, in fact had his own business. He was quiet, not the talkative type and liked to work alone. Since he worked outside his face always had color but on some visits the redness in his cheeks far exceeded the color of the rest of his skin. And on those visits his BP was usually high.

We saw him initially back in 2005 because he had no health insurance. He was a loner and his business was small, just him. Never went to a doctor actually. He felt well, worked hard all day and just didn't see the need of having a doctor or insurance. His family had lived in the area for generations, working in various trade services. Once our clinic was up and running he somehow found us. It might have been his wife who cajoled him into coming.

On that first visit his blood pressure was very high.

"Wow, doc, really? I feel fine."

High blood pressure can be silent until it eventually damages your heart and kidneys, or causes a stroke.

So, we got him on medications, talked about diet, salt and beer. His usual routine was to work all day in the sun, come home, drink a couple of six packs and fall asleep in front of the TV. Gradually over a year, we got him to understand that all of that beer was a factor in his high blood pressure. And for months he would cut down. His BP would be better and we would reduce his medications. We would talk about some of the landscape projects he did around town. He did some of his work on big estates. I would sometimes ask why he didn't hire some help. He would say it was too much bother and too much paper work. I think he didn't want to have to deal with anyone else on the job.

On one visit he talked about a job he was doing in Stockbridge. As he described the property and his dealings with the owner I realized that I knew the party involved. He told me about planting a ten foot pear tree where the owner had specified. At the end of the day when the tree was in and watered, the owner changed her mind. Didn't like how it looked against the background.

He had to take it out, move it and replant it. Knowing the owner I told him I was not surprised and smiled knowingly. On subsequent visits we would often kibitz about the landscaping desires of a certain property owner in the county. It usually brought a smile to his face, which I considered an accomplishment.

On some visits he slipped backwards in his habits. The beer intake grew. And on those visits his blood pressure was higher. I could tell even before the cuff was on his arm. His face was redder, the vessels more dilated, the pressure coursing through them higher. Like a barometer changing pressure before a storm his face color predicted the turbulence in his blood vessels, even before we measured it.

After Massachusetts passed its health insurance law, he gained access to insurance and private care. Once again I suspect it was his wife who goaded him into following through on this. He left our clinic and moved on to a primary care medical group that attends to his health needs. Now I no longer get updates on the landscaping escapades of a certain acquaintance in the county. I wonder if he still has the patience to work there.

Chance Encounter

The gesture puzzles, but has charm. A tall black man with round face and short-cropped hair turns his head slightly with a smile that he hides, shyly, behind an upraised hand. Not looking at me, he directs his attention on the patient at my side as we leave the check-out nursing room.

"Ah, you. You said you had an appointment."

I look from one to the other. These patients know each other. There is a friendly playfulness between them. And now they each know another thing they have in common.

The patient at my side is Albanian and works as a chef in a restaurant. His head is shaved bald. His body is stocky, tough looking. But his eyes convey intelligence and warmth. We had just finished his visit and were arranging a follow- up appointment and his prescription for a blood pressure medication. His English is good but accented. Accented how? Greek, Albanian? It could be French to my tin ear. When he responds to certain queries on his health his shrug is certainly Gallic. Early in his visit he was proud to inform me that he was a professional driver in his home country. I ask why not drive here? He replies that it is easier to get a job as a cook.

Finished with the Gallic seeming Albanian, I usher the tall man with the shy smile into the exam room.

"You two know each other."

"Yes, I work in the kitchen at the restaurant where he cooks." This accent is definitely French.

"And where are you from? What is your home country?"

"Ivory Coast. I have been here two years." His smile, no longer shy, now dazzles.

He is older than his chef friend, tall and lean but beginning to add an American paunch. And on a past encounter a colleague found his cholesterol levels to be American middle-age in range.

His exam is normal. We discuss cholesterol, its risks to him. I tell him I do not think he needs cholesterol lowering medication. On that past visit my colleague had pushed him to exercise more.

"Yes, yes. Now I walk one hour each night after work, even in the dark. They told me this. I do not want the American heart attack."

He tells me he eats all his meals at the restaurant because he works such long hours. He has to eat what they provide. It is not necessarily the healthiest fare.

I mention his friend the chef, the earlier patient he had greeted with his mock peek-a-boo gesture. His dazzling smile grows wide. I do not have to elaborate.

"Yes, yes. I will ask him. He will help me."

Two immigrants, different countries, different cultures, different skin colors. They will be on a mission now, helping each other get healthier in their adopted country.

Seven Years Later

Their stories get pieced together over years. Sometimes I remember parts, other times they remind me. Last week I saw a man for a preventive check-up. Prevention and health maintenance are promoted at the clinic. As a rule we are quite successful in this endeavor. When I usher him into the exam room I remember the face and the name but not much else. I have not seen him in two years.

He is from Guatemala, works as a landscaper here in the Berkshires. He feels well, has no complaints and has bought into our philosophy of getting a health check up every one to two years. He is in his late 30s. I scroll back through his records and note some elevated lipid levels which could put him at risk for cardiovascular issues down the line. I see that he has been admonished about his diet and alcohol intake in the past which resulted in improvement in those blood fat levels. His most recent numbers just before the current visit are not good, significantly higher than a year ago.

And sure enough, he has slipped back into bad dietary habits. We review the changes he has to make. He already knows them and promises to follow through.

We complete the physical exam. Everything is normal. As we finish up he asks me if I remember when he first came to the clinic seven years ago. I don't. I did not go that far back in his chart.

"You found a rectal fissure."

It comes back in a flash. He had been in the country less than a year. He came to the clinic the first time because of a rectal complaint. Before seeing a patient we are handed an intake form with the reason for the visit. Even though trained not to jump to conclusions, your mind automatically begins to consider possible diagnoses for the complaint even before you start questioning or examining the patient. My first thoughts on this gentleman were hemorrhoids. Common things are the most likely diagnoses.

He was not fluent in English back then. Through an interpreter he tried to explain a "pimple" next to his rectum that sometimes leaked fluid. Well, pimples are not uncommon on the buttocks and sometimes as pustules they can burst and leak fluid.

It was more complicated, however. The leakage of fluid occurred when he had a bowel movement. On exam it was not a pimple but an opening of rectal tissue completely separate from the actual rectum; a rectal fissure. Rectal fissures can lead to serious infections. They require surgical correction. We arranged this. It got done up in Pittsfield. All had been well since the surgery.

I thank him for reminding me and tell him I am glad everything still worked well. We have a few minutes before my next patient.

"Is your family still in Guatemala?"

"Yes. My wife and daughter."

"Do you get back to see them?"

"I went once three years ago. It's hard to travel home and then get back into this country."

"Do you ever think about getting a green card or citizenship?"

"I do, but it's really hard."

"How about going back?"

"I tried to start a business back home but I can make better money here."

He works for a landscape company five days a week, and then works week-ends on his own doing odd landscaping jobs. He sends money to Guatemala to support his family.

I am curious about his route from Guatemala to the Berkshires. He first went to Boston by ship. Since he had a friend/contact here in the Berkshires he made his way out here and started working. We have several patients from Guatemala so there clearly is a network of sorts. You have to admire what obstacles people overcome to try to make a better life for themselves and family. When I get the full stories over the years my admiration grows.

No Perfect Storm

Many patients who find our clinic for a health care issue are in transition. One such patient was a man from the west coast who had just become a widower. Jobless, adrift without his longtime wife, he made his way to the Berkshires where he had family.

This man arrived with serious cardiac issues. Ten years ago he'd had a heart attack and subsequent to that underwent cardiac bypass surgery. Without proper medication and attention to his diet he sometimes slipped into congestive heart failure.

Here in the Berkshires he had a place to stay and family to look in on him, but he was often alone, without friends or anything substantive to do. He got depressed and anxious, started using alcohol and became irregular with his medications. An ER visit and hospitalization resulted. Without Massachusetts health insurance and without a doctor, he got referred to our clinic post his hospital stay.

When I first encountered him I was impressed by his knowledge of his medical history. He kept a folder and was able to relate the complicated details of past procedures and hospitalizations with accuracy. I was also impressed by his examination. He was a solid bulk. The leading edge of that bulk was a protruding abdomen. Some men gain this body form with age. Usually, most of it is fat, but in him it was solid muscular-like fat. In addition to the surgical scar on his sternum from his bypass procedure, he had many other scars on his hands, arms and legs. He also had serious venous varicosities in his lower legs.

I was curious about all the scars. They were not all from surgeries. Fights, I wonder? He does not admit to that but does relate that he was a chef on a tuna fishing boat out of San Diego for many years and often cut himself with the kitchen knives. Chopping food on the high seas had some additional risks, I guessed.

On a follow up visit, once he had stabilized, we had a little time so I asked about some of his adventures on the fishing boat. I wondered if there were times his boat was in trouble in a storm? He eyed me with disdain. With a good captain and good instruments you should never get in trouble, he said. And his captain had always been good. Perhaps the only big storm he encountered was the one where his bad heart and a sea of loneliness threatened to sink him.

The Diastolic Rumble

The physical examination of the patient is a dying art. Modern technologies like x-rays, CT scans, MRI scans, sonograms, and blood test analyses all reveal many details about the body. This is all well and good – the more information we have, the more accurate the diagnosis. But all that modern technology costs money. Seeing patients in a free primary care clinic for those without insurance and access to care requires that we look at, listen to and examine patients before sending them willy-nilly for tests and scans. Getting such things paid for takes effort. Plus, some of us still pride ourselves on making diagnoses the old fashioned way.

Not infrequently, people enter the clinic never having been examined by a doctor before. A few years ago a young Guatemalan woman sought attention because she and her husband wanted to have a baby. She wondered if she was healthy because she seemed to get out of breath easily with activities. She wasn't healthy. As soon as I put my stethoscope over her heart I knew she had serious issues. She had two loud heart murmurs. Subsequent evaluation revealed complex congenital heart disease requiring surgical correction. If she'd had access to medical care as a child it would have been corrected long ago.

She went on to bear a child, tolerating the pregnancy and birth without complications. She returns for her visits with healthy son in tow. This summer marks the 6th year since her surgery. As part of that surgery she required implantation of a prosthetic heart valve. For medical and child bearing purposes that valve had to be a tissue, not a mechanical prosthesis. Mechanical valves require anticoagulation with drugs that can cause fetal malformations. Women cannot get pregnant on such drugs. A drawback of tissue valves is their longevity. They usually function well only for about ten years, a significantly shorter lifespan than mechanical valves. Unfortunately, our patient's tissue valve lasted just six. The leaflets wore down and became calcified. The valve became nearly immobile, obstructing the flow of blood from left atrium to left ventricle. The medical term is mitral stenosis. It results in symptoms like shortness of breath and leads to congestive heart failure.

The problem usually requires repeat surgical correction. She had been watched over the years both by our clinic and by cardiologists in Worcester, Massachusetts. In the past year we knew trouble was brewing. Yes, that knowledge came primarily through imaging technology but it also announced itself on the physical examination.

As a cardiology fellow/trainee in the 1970s a rite of passage was making the diagnosis of valvular heart disease with just a stethoscope. One of the more difficult disorders to appreciate is mitral stenosis. It causes turbulence of blood flow in diastole when the heart is relaxed. The resultant murmur is subtle, low in pitch and rumbling in character. It is a little like a distant roar of ocean surf miles away.

And mitral stenosis has become quite rare in the U.S. because strep throats in childhood are treated. As a result a sequela of strep, Rheumatic Heart Disease, is very uncommon. Our patient did not have stenosis before her surgery. It was a result of the unusual wear on her tissue prosthetic valve.

So when I heard this murmur early this summer old teaching instincts were awakened. I wanted to look around for medical students and residents. I knew none were available but I had to find somebody. I corralled Ilana, our Director of Medical Services. She is a Nurse Practitioner.

"Listen to this, listen to this! It's very unusual to hear a mitral stenosis murmur."

Our young mother from Guatemala has had repeat surgery to repair the problematic mitral valve, and she is doing fine.

A rumbling murmur in the heart of a young woman warmed the cockles of an aging diagnostician's heart.

A Little Here, a Little There

The voice comes from behind as I walk toward my car.

"Hey, Dr. Blake, Dr. Blake."

Not many people around town know me as a doctor. Friends and acquaintances would have called out my first name. I know from my days as a practicing doctor that being hailed in public means that a patient has seen me and probably has a question. My mind clicks from relaxed retired to professional alert.

The hearty voice is familiar and I recognize its owner as soon as I turn around. He is a big man with a big smile. He also has a big blood pressure problem. Glad to see him, I walk back to where he stands on the steps of the Post Office. He'd missed his appointment last week and I was concerned. Like most of our patients he is a paid hourly worker with no health insurance. If he doesn't work, he doesn't get paid. Because of this issue we have evening hours one day a week. For several months his appointments were scheduled after five pm so he could come in after work. But he missed last week, and as it turns out, that is what he wants to talk about.

"Sorry I missed the appointment. The clinic did not call to remind me and I forgot."

I am not sure about the reminder call. Our staff usually does that regularly. I ask if he is okay and if he has enough medication to last him until next week when I will be back in the clinic. He says he is fine on both counts.

He had a previous doctor treating his blood pressure but had lost his health insurance and could not afford to continue seeing the doctor and pay for his medications. Sadly, this is a common occurrence, even in a state that has mechanisms to provide health insurance to those without it. He knew enough about the potential problems from high blood pressure that he found his way through our doors late one afternoon early in the summer. We found that he maintained swimming pools for his employer a lot better than he did his own health. He smoked cigarettes, ate all the wrong things, and never checked his pressure at home.

Over two months we made good progress. He agreed to attend a smoking cessation program, worked on his diet, and we got him back on some medication he could afford. His blood pressure improved but was still not in optimal range. When he missed his appointment last week I worried that we might have lost him. Things come up for those living on the edge, just scraping by. I have learned that we cannot always understand the decisions they make about their health care.

"I'm in the clinic next Tuesday and Wednesday. Call later today and make an appointment. Or just come by if you cannot get through."

The big smile lights his face. He grabs my hand.

"Thanks, Doc. See you then."

I had felt from our initial meeting that he was committed to controlling the blood pressure issue. But I am never sure. Getting some folks to buy into their health care is often a big step. People face obstacles, both real and imagined. I hoped he would show up next week.

A cold breeze hits me as I walk back to my car. I tug my hat down low and stuff my hands into the pockets of my fleece jacket. Julie is waiting in the car. I climb in. She says nothing but she probably notices the small smile playing on my face. You have to take these brief little victories when you can.

The Pain Game

The intake form said "New patient. 48 year old woman. Back pain, chronic. Needs more meds. Has records."

I groaned inwardly. "Needs more meds" raised red flags from my past experience file. I had never dealt well with chronic pain patients on various prescription pain medications. There were so many layers involved and I usually did not have the patience to peel through them. My general approach was to try to coax them off pain killers. My track record lacked success.

I gritted my teeth and opened the door. She sat on the exam table holding a sheaf of records. "Hi, Mrs. X, I'm Doctor Blake."

"Hello, Doctor. Would you like to look at my records?" Nice voice and nice smile.

"Yes, I would, but how about we talk a little first. What brought you here today? How did you find us?"

Lack of medical insurance, no primary care doctor, and limited financial means are criteria to get past the clinic's initial vetting process. At first glance this lady did not have the look of someone with those attributes. And considering the thickness of the file in her hand she had seen plenty of doctors.

"I'm new to the area and am running out of medication. I have very bad back pain. I need your help." The voice still nice, but carrying notes of supplication that jarred me a bit.

I went through the process of establishing the duration of her problem (5 years), the apparent cause (a car accident), any surgery (yes), current meds (many, including oxycodone), and her current pain level (severe.) Plus, she had high blood pressure that required two medications for control. I felt some relief. At least this was something I might be able to help her with.

"Okay. I need to examine you. I'll step out while you change into a gown. I'll review your records while you change."

The records included surgical procedures, CT scans of the lumbar spine, visits to orthopedic doctors, and two different primary care doctors. Her meds included not only the BP meds but oxycodone, valium, and an antidepressant. It was a wonder she could stand up. The back injury was substantial, two fractured vertebrae requiring surgical stabilization. I did not see any records of follow up physical therapy or mental health counselling.

I went back in and established that she had been recently divorced, had to move away from her home, left her doctor fairly abruptly, and was running out of pain meds. Tears now.

Her blood pressure was high but her carotids, lungs and heart all sounded fine. I asked where the pain was most severe. I stood just to her left side. Most patients would point or state where the pain was most bothersome. She grabbed my hand and placed it on her lumbar and left hip area. I was taken aback, raising my eyebrows at the medical assistant to her right.

I said, "OK. Please relax and let me examine the area."

Alarms were pinging in my head. I put her through some range of motion maneuvers, palpated for tenderness and deformity, and checked her lower extremity deep tendon reflexes. She winced a little with one of the maneuvers but other than the surgical scars and mild tenderness nothing else seemed out of the ordinary.

"Well, okay. For sure we need to renew your blood pressure meds. Are you taking them all regularly?" She said she was. I was not so sure.

"Regarding your back, I am not sure how much we can do. I do not detect any red flags or increased spasm. I am concerned about the amount of narcotic meds you are taking, particularly in combination with the BP meds. I would like to try to reduce them gradually."

"Oh, no. I can't do that. I've tried. And with all the stress of my divorce…. Well, everything is just worse."

Not sure what to offer I look at the patient and then at the medical assistant. I had no doubt this lady was hurting in many ways, but she wanted big time amounts of narcotic pain pills and my antenna were detecting elements of drug seeking behavior. What to do?? I decide to stall and compromise.

With a sigh and what I hoped was a friendly shrug I look back at the patient.

"Well, I cannot give you the large amounts of narcotic pain medication you request, particularly with your two blood pressure medicines. It is not safe." That was true, but it was also true was that if she really was in a great degree of pain that could drive up her blood pressure.

"What I can do is give you a two week supply of your pain medication at the same doses, and have you come back. I want to see how your BP responds to resuming those medicines. We should also talk about some mental health counselling. That can help with chronic pain and stress."

Her manner and voice reflect clear disappointment. "Is that all you can do? I've tried counselling. It doesn't help."

I shake my head and say, "Sorry. Please get dressed. Lynne will have your prescriptions and appointment for you outside. See you in two weeks."

I excused myself, and left the room a bit frazzled. I take pride in my ability to forge good relationships with patients. I had failed here. My perception of her attempts at manipulation had put me off. I would treat her blood pressure but would need help in handling her pain issues. My mind searched for possible referral options as I stood outside. Lynne, our Clinical Care Coordinator, approached. I rolled my eyes and told her the plan. I also asked her to talk with the lady about a possible mental health appointment.

In two weeks the patient called an hour before her scheduled appointment saying she could not make it. She had asked Lynne to have me renew her current narcotic pain prescription for another two weeks. Lynne had tried to tell her we could not do that. She came to me saying the patient would not take her response. I didn't want to, but took the call and re-iterated my position to the patient. We could not refill her narcotic meds again. It was not safe. I asked her to make another appointment. She refused and hung up. We never saw her again.

This encounter took place eight years ago. This lady clearly needed help with her dependence on narcotic pain medications. We had failed her. Our small free care clinic was not set up to offer all the attention involved with pain medication dependence issues. We still aren't, but we now do have more options for treating chronic pain. Maybe offering this lady acupuncture, massage, and physical therapy, as well as psychological counselling, would help her today. I still wonder how it all turned out for her. I hope she got help somewhere.

New Golf Pro

It sounds paradoxical but it takes a lot of money to run a free clinic. Yes, we have many volunteers providing services but they require the support of staff, office space and equipment to do their clinical thing. Paying for staff salaries, rent, equipment, supplies, and utilities adds up to a pretty penny. We need to raise a lot of money to keep the doors open.

All that money comes from private donors and some foundation grants. Last year, in a one- time occurrence, a sizable amount came from a golf tournament. Non-profit organizations sometimes use golf tournaments to raise money. I have played in a few. They can be fun events, give a nonprofit visibility and often raise modest sums, but it is unusual for them to raise substantial amounts of money. There is considerable expense in staging these things, in addition to the time and effort of those who put it together.

Last year our clinic was the chosen beneficiary of a tournament put on by Blue Cross/Blue Shield of Massachusetts. They choose a different non-profit each year. We had somehow crossed their radar screen and got the nod. All that was required on our part was to put together a foursome, show up, play and collect a check. No organizational expense, no staff and volunteer time commitment. What's not to like?

Golf is a very difficult game, a good walk spoiled in some people's minds. It requires so many integrated movements. The ball is struck by clubs of varying shape and length, travels on uneven terrain and rests in different types of grass - fairway, rough, green. Sometimes it even lands in places you where you can't find it, like water or dense woods. Yet it seems so simple. The little white ball sits there, nice and still, just waiting for your calculated strike to move it to your desired target. When something is very hard to do, but seems like it should be easy, the human mind gets frustrated at less than optimal results. Hence, the mental aspect of the game – keeping oneself calm and accepting those undesired results.

Like most amateurs who struggle with the game I marvel at the professionals I watch on TV. The results they accomplish are truly remarkable. They get paid handsomely to play a game. Well, I have now joined their ranks, at least the monetary part of it. I played in a golf tournament and got paid.

The day was cool, blustery. The course was hard, a definite challenge. Our foursome staggered around but did manage some good shots, at least from the amateur perspective. Just like the pros at the end of the round, we signed our scorecard, posted it with the tournament director and awaited our spoils. Unlike the real pros, we had won even before we started. Our score meant nothing. We knew we were going to get first prize money.

Typically, these events have a dinner and awards ceremony following the round. I knew we would be asked to say a few words of thanks and a bit about our clinic. I was prepared to do that. Just before the ceremony the tournament director informed us of how much money they had raised. Having participated in other similar events (not as a beneficiary) I expected at most $20-25,000. *They raised $75,000!* I was blown away. It was all going to our clinic. When I spoke minutes later I had trouble maintaining composure due to the emotional surprise at the amount of money donated to our small clinic. I hope I managed to convey genuine gratitude. And the day had a bonus. I got to play in a golf tournament and go home with a big check, just like a pro.

A Romantic Streak

Many of our Latino patients speak a fair amount of English. But often they get lost when a query is long or complex. So we usually have an interpreter present. The patients will respond to basic questions in English, but when they wish for clarification on the question, or need to elaborate on their response, they shift to the interpreter. The current patient and I go back three years. I like his sense of humor. We are conducting a "review of systems" as part of his annual preventive exam. We also see him for hypothyroidism and high cholesterol. The purpose of a review of systems is to draw out things that are bothering a patient and may affect long term health. The systems reviewed are heart, lungs, gastrointestinal, urinary, etc. I have moved into questions pertinent to mental health.

"Anything making you feel sad or depressed?

"Not sad really, but I cry for love." This exchange is in English.

He says this with a smile and twinkling eyes. I have a feeling this is not at all about being depressed. He had often expressed the joy he found in his relationships with women. The interpreter is amused. I can see it on her face.

Remaining in professional mode I follow his "cry for love" response with, "Are you having problems with your girlfriend?"

"No, no. It's wonderful. Love makes me cry." Again, the smile and twinkling eyes.

He is in his early 50s and works as a gardener. He needs medication for his thyroid condition and high cholesterol. He tries to lose weight but struggles on that front. His love life is clearly a source of joy. On a prior visit he expressed some concern with erectile dysfunction. There was nothing on his examination that suggested a physical cause. I had mentioned stress or a psychologic issue. He did not buy that and subsequently solved the problem with an imported aphrodisiac from his home country of Colombia. I ask about his current sexual function and he confirms that all systems are go.

We conclude the visit with refills of his medications, schedule a nutritionist appointment, and a screening test for colon cancer. Much of this is now in Spanish. We stand and move toward the door. Dropping my professional demeanor a notch, I ask if he has more than one girlfriend. His eyes and smile grow wide. He moves his index finger to his lips and mouths a silent shh.... I meet the interpreter's eyes. They smile as well. She, like me, is amused by his infectious good humor.

Doo Rag Man

A doo rag serves two purposes for a bicyclist. It absorbs sweat and keeps the sun off tender scalps devoid of hair. Cycling helmets are designed with multiple vents. Failure to apply sun screen and/or one's doo rag can result in weirdly patterned sunburn after a day's ride. I am a devoted wearer of such attire.

Given my biking interest, I also appreciate the style opportunity doo rags offer. So when a patient appears attired in red checkered rag neatly knotted at the back I am intrigued. He is a cyclist, but his wheels are of the motor powered variety. He seeks medical attention for a painful shoulder. He fell more than a month ago and the shoulder still bothers him.

Before getting to the shoulder I notice his vital signs – blood pressure, pulse rate, and body weight. The blood pressure is disturbingly high. That leads to careful questioning about his past history, previous medical check-ups, and treatment of the blood pressure. He is in his mid-50's.

"Never much bothered with doctors, doc. Always felt pretty good 'til I fell on this damn shoulder."

The shoulder is bad. The range of motion is severely limited and he has point tenderness near the rotator cuff tendon attachments. Considering the month of symptoms he has probably torn some of the tendon fibers. We devise a program to deal with the shoulder. I then get back to the blood pressure. He needs to make life style adjustments, get some blood tests and start taking medications. Grudgingly he agrees to all that, at least the lab tests and medications.

That summer we see each other several times. He always wears a checkered doo rag, ranging in color from red to black to blue. I tell him of my use on bicycle rides but admit to their monotone drab colors. His blood pressure improves, not to ideal levels but enough to take him out of dangerous territory. The shoulder does not get better with conservative treatment and physical therapy. He will need surgical correction if we can get it arranged and he agrees to it.

On one follow up visit he asked "You ever ride a motorcycle, doc?"

I had not.

"You should take a look at my Harley out front."

I do. It's a big, mean looking machine. Not anything I would want to ride. I will stick to my leg powered bicycle. I wouldn't mind having one of his checkered doo rags, however.

Echoes in Miami

I am about to see a patient for a follow up visit regarding some worrisome chest symptoms. I had ordered an exercise echocardiogram three weeks before. The medical assistant hands me the printed report of the test. The study is dated the week prior. I notice the address of the cardiology office submitting the report – Miami, Florida. This does not compute. Our clinic, the patient, and I are all in Great Barrington, Massachusetts.

I reboot and read the report. It's good news. The patient's exercise tolerance is fine, the muscle function of the heart is normal both at rest and at the conclusion of the stress/exercise. Plus, all four valves work normally. This means the symptoms I am concerned about are not cardiac in origin. We can focus on more benign issues like her weight and exercise conditioning.

The patient enters the exam room. She is anxious about the test results.

"Please, doctor. How is my test?"

"Good news. Your heart looks fine both at rest and with the exercise. I do not think your chest symptoms are related to your heart."

She is pleased, visibly relaxing into her friendly smile.

"Now, I'm confused about this echo test. The cardiology office where it was done is in Miami. Is this right?

"Yes, doctor. I had the test in Miami."

"So, wait a minute. You are working here in Massachusetts. How did you get to Miami and back, and how did you schedule the test down there?"

We refer our patients to Fairview Hospital in Great Barrington for most tests, including the stress echoes. Massachusetts Health Safety Net pays the hospital for the test in situations where the patient is uninsured. The cardiologist who interprets the test generally waives the professional fee for our uninsured patients. Our patient in question works as a housekeeper for a family that summers here in the Berkshires, but resides in Miami. The patient is not a resident of Massachusetts. She cannot get the MA state supported free health care insurance, or any form of Obamacare.

The Fairview Hospital charge for the stress echo test would have been $950. The patient did some homework. She could get the test at a cardiology office in Miami for $250. She found a late night "fly and return" airline ticket to Miami for $175. She also did the math. Her housekeeper income could better tolerate the $425 versus the $950. So she flew to Miami on a cheap flight at 11 pm, slept a little in route, had the echo in

Miami at 10 am the next day, and then flew back to Hartford, Connecticut that afternoon. I am astounded at her ingenuity. I can't bring myself to ask about transport to and from the airports, but I suspect she worked that out economically as well.

The patient relates her story with a smile and her eyes twinkle at my astonishment. The immigrant population that we serve has to be resourceful. I continue to be enlightened and heartened when I encounter such examples. Most patients are matter of fact about it. This is just something they have to do in certain situations.

Hormones Make the Man

A young Latino man presents to the clinic because his nose bleeds. He is seen by Ilana Steinhauer, our Director of Medical Services/Nurse Practitioner. She stems the bleeding, suggests ways to deal with future bleeds. The young man is 22, recently in the States from Mexico, and works as a restaurant kitchen aide. Beside the nose bleed, a few things about him bother her. She notices he speaks in a high voice and has very little body hair. She wonders about a hormone issue and schedules him to see me during my hours a week later.

When I see him it is clear something is not normal. He speaks in a pre-pubertal child-like voice and has no visible facial or body hair. The answer jumps out when I examine his genitalia. There is no pubic hair, his penis and testicles are small – those of a child. This man is hypo-gonadal. At age 22 he has not gone through puberty. His body has not secreted the hormones, mainly testosterone, to make the normal age-related body changes. The question is why?

He is of normal height and weight, but thin, not well muscled. I ask about girlfriends and sexual activity. He has had neither. Is he worried about all this? Yes, he would like to know why he is different.

I step out of the exam room and discuss the situation with Ilana. She had suspected as much, but thought he would react better to a male examiner with questions on such sensitive issues. We consider what to do. We could start with a battery of blood tests to check basic function and hormone levels. Since he will need to see an endocrinologist eventually, we decide the best course is a direct Endocrine referral. They can decide how best to proceed.

The endocrine work up confirms our diagnosis. He is hypo-gonadal and will need hormone supplementation. But to our surprise the cause of hormone deficiency is a marble- sized cyst in his pituitary gland. The pituitary is the master gland; it controls all other endocrine gland function such as thyroid and adrenal. His cyst has compromised the normal pituitary tissue to such a degree it cannot function adequately. The cyst has to come out. This requires delicate neurosurgery. How and where do we arrange this?

The endocrinology consultant suggests a surgeon in Worcester, Massachusetts. We can arrange that but the bigger issue is getting this shy man, without a car or a driver's license, over there for pre-op evaluations, then surgery, then the post-op visits. It's a two hour ride. The clinic, VIM, seems to rise to these occasions. Ilana, our Director of Medical Services, and Gladis, our

Medical Assistant, get on these issues like bees on pollen. I am constantly amazed at what they accomplish to get what patients need. A volunteer driver, with fluent Spanish, is found to make the trips. Our young man becomes comfortable with the driver. They become friends. The surgery goes well. The cyst is removed without complication. The patient is now receiving hormone treatment by the endocrine consultant. His body is becoming that of a young man. His confidence grows. When last observed he was talkative and smiling, things I had not seen on his pre-op visits.

Hats off to our volunteer driver. He spent the large part of several days with our young man and clearly played a big role in his health care.

During my twelve years as a VIM volunteer I have frequently been surprised and amazed by the undiagnosed medical issues that walk through our doors. I had learned about hypo-gonadism some 40 years ago in medical school, but had never seen an actual case. Now I have. One of the joys of medicine is learning something new from each patient you encounter. Thanks to VIM my education continues.

Hard to Breathe

We all get out of breath. A quick climb up a flight of stairs with a bag of groceries, a short sprint to catch a bus, or a brisk hike up a steep trail leaves most of us panting. But we recover. And the act of breathing retreats to its unconscious, metronome-like regularity.

Sure, as we age or lose fitness that recovery may take longer. We don't give it much thought. It's not hard to breathe. We're just a little out of shape. But when breathlessness lingers for several minutes after stopping an activity, or comes with just a short walk to the refrigerator, we might get worried and seek an evaluation.

The man sitting across from me looks healthy. His ruddy complexion and stocky musculature suggest outdoor activity and the strength to tackle most things. But he is worried about his breathing. Two weeks ago he swam to shore from the sailboat on which he crewed and almost didn't make it for lack of breath. Plus he had to sit on the beach for five minutes before he could get up and walk to the marina.

I run through a series of questions. Chest pain? No.

Past history of asthma, heart or lung problems? No.

Smoker? No. Use of beer and alcohol. Sure.

Blood pressure high? Not sure, never really felt the need to see doctors.

He is a nomad, having traveled the world on various sailing regattas working as crew. He has also taught sailing at clubs and resorts around the country. Currently he is between trips and jobs. He has no health insurance.

What is he doing in the Berkshires? We are a long way from the open ocean. Well, he finished his last sailing gig two weeks ago on the Connecticut shore of the Long Island Sound. He is visiting a sister in Stockbridge he hasn't seen in two years. When he mentioned his breathing issue she told him of our clinic for uninsured patients.

"So here I am, doc. What do you think?"

He is 52, has thinning red hair, and although muscular is definitely over weight. And his blood pressure is high. It could just be the weight and blood pressure. But when I listen to his lungs I hear crackling sounds with each inspiration. And feeling his chest to check his heart size the forceful beat is pushed laterally to the left side of his chest. His lower legs are swollen. The crackles mean fluid in his lungs, the displaced heart impulse means an enlarged heart, and the swollen legs are edema - retained fluid. It adds up to congestive heart failure. His EKG confirms the enlarged heart. Fortunately it gives no evidence of a prior heart attack.

I explain why he gets out of breath, what the diagnosis means, and that he needs additional tests and treatment. He buys it all and admits the breathing symptoms had been brewing for a month or two. He allows that the 50 yard swim to shore two weeks ago really scared him. "Thought I was going to die, doc."

We see each other every two weeks over the summer. His symptoms improve, his BP comes down and the fluid in his lungs goes away. Additional testing does not reveal evidence for blocked coronary arteries but confirms the enlarged, weakened heart muscle. We cannot cure that but can help it compensate and prevent further deterioration. Most likely the heart muscle damage was from unrecognized, untreated high blood pressure for several years. His use of alcohol may also be a factor.

In the fall we help him get health insurance and connect him to a private cardiologist in Pittsfield. He moves on, hopefully to several years of stable health. This visiting sailor had no idea we existed or how he could get medical attention when he arrived in Stockbridge. But his sister did. The word is out.

The New Entrepreneur

Our windows need washing. A quick perusal of the Shopper's Guide yields three numbers. On the first two, I leave voice messages. Option three answers. The accent is Latino but his English is fine. He asks questions. How big is our house, how many floors? We talk price. It sounds reasonable. I ask when he can do the job.

"I can be there with my crew on Monday."

Monday dawns cool and cloudy. The pick-up truck rolls in the driveway around 9 am. Four men pile out with squeegees, buckets, and rags. The driver approaches me. He looks familiar.

"I'm Freddie. This is my crew. Do you have an outdoor water spigot?"

I definitely recognize two of the crew. They smile shyly.

They both had been in the clinic within the last two months; one for a bad case of poison ivy, the other because of an abscessed tooth. I had seen the man with the bad tooth for a general medical exam. Our dentists prefer that we medical volunteers check out patients who have not had access to health care before they do any major dental work. Many of our Latino patients have never seen a doctor. Sometimes we uncover untreated medical issues. This man was fine and the abscessed tooth came out without issues. The other man's poison ivy had cleared with a course of steroids.

I walk Freddie out back to show him the water hose. I suspect that I may have seen him in the clinic in years past but am not sure. I say nothing.

They are done in two hours. I tour the house looking at the windows. They sparkle. Freddie and I settle up by his truck. I tell him he looks familiar and I know two of his crew from the clinic.

"Yeah, I saw you at VIM four years ago. I hurt my back. I got insurance now. My guys are new in the U.S. this summer.

In addition to health insurance he also has a business now. Driving around town you just might see his sign. When he is working on a property he plants one of those mobile advertising signs – "Freddie's Window Washing."

This is a story I have seen numerous times over my twelve years at the clinic. An immigrant arrives, has a health issue, and finds his way to our clinic. Over a few years he works hard, becomes a citizen and gains access to the U.S. health care system. Often he becomes a cog in a network of similar people. And sometimes he starts a business that helps other new people from his homeland get a leg up here. I think we call this America. No walls needed.

The Cook

Some patients look sick at first glance. The man entering the exam room fits that bill. Slow gait, disheveled hair, dour expression, sunken eyes and sallow complexion all speak to illness. He is skinny and reeks of cigarette smoke. If he were a dog you would say he was on his last legs.

Once he sits he meets my eyes. I half expect to see jaundice there but do not. I see despair. He is in his early 60's, just lost his job as a cook at a restaurant, and needs medication for high blood pressure. He lives alone, no spouse, no family in the area. His BP has been a long time issue. No job, no insurance, no hope. But he found his way to VIM. Maybe we can help him.

His blood pressure is high, even on the meds he is taking. His lungs are full of cigarette related gunk, his heart enlarged and overworked. I also detect circulation issues in his abdomen and legs. We have our hands full.

That first summer we make a little progress with adjustment of his meds. But he still smokes and remains depressed. He continues to look lousy. He is not interested in mental health counselling.

That fall we decide to enter him in a new program called Shared Medical Appointments. Thanks to Ilana Steinhauer, our Director of Medical Services,

VIM Berkshires was asked to get involved in the program and received a grant to set it up and run it. The concept is to have group sessions of several patients with the same diagnosis. The selected patients all meet together for a 60 to 90 minute session. In that extended session they receive body relaxation exercises, share common issues and concerns about the disease, and receive education and dietary advice. All this is in addition to a quick individual exam and adjustments to their specific treatment plan. The clinic had chosen hypertension (high blood pressure) as the first diagnostic grouping for the program. Shared medical appointments mix medical and group-psychology therapy. In addition to the clinical provider, the patients interact with a group facilitator who runs the balance of the sessions. Our man, the cook, joins five other hypertensive patients. I am somewhat surprised he agrees to sign on. The group meets monthly for six months with occasional additional sessions in between.

I am away for several months. I return late the next spring. The transformation is astounding. In contrast to our first encounter the cook now looks engaged, almost happy. He smiles at my greeting; his eyes show life, not despair. His blood pressure is fine. He has stopped smoking. His lungs are improved. He has a new job cooking at a local school.

An advantage of a small, independent clinic like VIM is flexibility. We can quickly add innovative programs like the Shared Medical Appointment concept. We can try them, tinker with them, and make them work for our clinic. It requires staff input for sure, but much is done by various volunteers. This particular program worked well for all six participants. They bonded. They improved. Life style and the mental approach to illness and disease are such key factors. The better patients incorporate those changes, the better they do. For patients like ours who have busy, stressful lives struggling to get by, these changes are hard to maintain. The SMA program helped our hypertensive group.

Our man the cook has other issues. We continue to work on them. His first visit is now over two years ago. Every time he comes in he reminds our Director of Medical Services, me, and anyone else he sees, that we saved his life, that we really cared about him and that we made him care about himself again. That's what it's all about.

Who's in Charge?

I'm fine with computers, I surf the net, I have upgraded to Windows 10. I have a smart phone, I text. Okay, I'm not on Facebook and I don't tweet, but I get by in this internet age. Yet this summer I have the eerie sensation that Hal has been resurrected from his *Space Odyssey* and invaded our clinic.

We recently installed Athena, a new Electronic Medical Record (EMR) system. For the past eleven years I cruised our previous system with ease, knew every little nook and cranny, got through patient encounters smoothly. Now a Greek goddess has landed in the clinic laptops. She has tipped the scales of medical encounter justice to the digital realm.

She's powerful, this goddess. Patients are tracked like prey in the forest as they move from "Check-In" to "Nurse Intake" to "Ready for exam" to "Check-Out." "Dr. Blake, your patient has been in the clinic 13 minutes." The screen is loaded with information, all available if you find the right tab to unlock Hal's tight fists. Chief complaint, present illness, past history, review of systems, vital signs, flow sheets, allergies, physical exam, orders, assessment and plan. It is all there in black and white if you peel back the layers.

"*Buenos Tardes, Senor X.* I am Doctor Blake." The patient and I meet. I know from the schedule he is here for a follow up of his hypertension and diabetes. I remember him, but not the details of his last visit three months ago.

"Just a sec. Let me look at your chart." I find the past history tab but when clicked it just lists some past diagnoses. Where is the note from the last visit? And what was his last blood sugar level?

"I'm sorry. Did you say something?" I am lost in the computer chart. I refocus. We proceed with the visit, me asking questions, engaged with the patient. I repeat his blood pressure; listen to his carotid arteries and his heart.

"Okay. Your pressure is fine today. No change to your meds. I need to find your last blood sugar result. And I hear a soft heart murmur. I will schedule some tests and have you come back to go over the results. See you then."

I usher him to the door with the interpreter. She will escort him to the check-out nurse. Now back to Athena. I have to enter a diagnosis, code it, enter a plan, and order the tests. No paper in this office. No conversation either. Just key-stroke the stuff in and the check-out nurse reads your finely wrought diagnosis and the orders on her computer screen. In this case it will be a while because I need to find the right place to enter my thoughts, make sure entries are properly coded. What the heck is the system diagnosis for heart murmur? What code will the Greek Lady accept?   I ponder, I search.

"Dr. Blake, your patient has been in the clinic 47 minutes."

"Yes, Hal. I know."

I am not sure what the patient was thinking but I am uncomfortable with the encounter. I was key-stroking and tab-hunting when I should have been looking at and listening to the patient. Yeah, I know. I will get better and faster as I learn the new system. And it *is* full of quality safeguard issues. But am I a better doctor? Is the patient better served? Some days I want to channel the *2001 Space Odyssey* astronaut, Dave, and unplug Hal… listen to him circle the drain as Mighty Athena's scale is tipped back in my favor.

Feeling Better

An elderly Latina limps into the exam room, greeting me with a crinkly smile. Her weathered brown face bears the effects of years in the sun. It's a warm face, friendly, one you can study. She has a grotesquely swollen right leg, but that is not the reason she limps. Today she is favoring her left side.

The right leg is quite a story. When first confronted with it eight years ago I was aghast. How did it get this way? We never did get official records. All we could establish via interpreters was that 20 years ago in Guatemala, a toe from her right foot had been amputated for suspected cancer. They had also dissected and removed all the lymph nodes in the upper right leg and groin. Without the proper lymph channels, lymphatic fluid accumulates, resulting in chronic massive lymphedema. And there's not a lot to do about it. Diuretics don't work. She drags that heavy leg around without complaint.

She and her family have been taught to massage it, wrap it, and keep it elevated. A few times over the years she has come in with areas of the leg reddened and oozing fluid. Cellulitis. Fortunately those infections respond to antibiotics.

We manage her high blood pressure and diabetes. Neither is too severe. Most visits are spent reviewing other minor concerns. Today her left knee hurts. I am not surprised because that knee is severely deformed from osteoarthritis. She also has a ringing, buzzing sound in her right ear. This is a new complaint. Tinnitus, most likely.

I ask various questions, listen to her interpreted responses, look in her ears, examine the left knee, check the tightly wrapped right leg, and make sure her heart and lungs sound fine. Her BP and diabetes are in range. I explain the reasons for her new symptoms; make a few tweaks in her regimen. Little things.

As we conclude the visit she is happy. *"O Dios mio. Me gusto. Me gusto."*

We both stand and she grasps my hand in both of hers, her face alight with her wonderful smile.

"Gracias, doctore. Gracias."

She limps from the room. I smile, marveling at her quiet dignity in the face of her ills. I am not sure I can tell you who benefits more from her visits – patient or doctor.

Notes:

These stories are based on my recollections. All quoted statements in the stories reflect my best recollection of what was said. The quotes are used to make the stories flow better.

As noted in the preface, the work done by VIM Berkshires volunteers like myself could not happen without the support of a dedicated staff and a committed Board of Directors. My sincere thanks to all of them. One director, Matt Mandel, graciously allowed use of his name in one of the stories. A tip of my hat, Matt.

Two current staff and one former staff person are mentioned in the stories. They too, graciously allowed the use of their names. I have worked closely with them over the years. They have my special appreciation. The former Clinical Care Coordinator, Lynne Shiels, was instrumental in building our patient population during the clinic's first several years. She devoted countless hours and extra effort to the clinic. Gladis Rave has been the Medical Assistant at VIM since we started. She is bilingual and serves as a medical interpreter as well. She has poured her heart and soul into the clinic. Lastly, Ilana Steinhauer, a Nurse Practitioner and our Director of Medical Services, has her finger on the pulse of everything happening at VIM. She is full of energy and passion. VIM's recent growth and the addition of new programs are due in large part to Ilana's efforts. She also sees patients on a regular basis and is one fine clinician.

VIM Berkshires is in good hands.